Dinosaurs Kingdom: For kids who really love dinosaur

- **Learn**
- **Color**
- **Draw**

<table><tr><td>

Learn

</td><td>

So, you love dinosaurs. You're probably familiar with the *T-Rex,* the *Triceratops,* and the *Stegosaur* — but you want more. Well, don't worry, cause we'll give you more.

</td></tr></table>

It would far too long to go through all the dinosaur species we know existed, so we'll have a look at some of the most representative dinosaurs, discuss what their name means, what they looked like, and add in a little bit of trivia. But first, let's look at how dinosaurs as a group got their name.

How dinosaurs got their name

The term was coined in 1842 by English biologist Sir Richard Owen. The name dinosaur literally means "terrible lizards", which we can safely say is a pretty fair description (at least in most cases). It comes from two Greek words: deinos, a name that means "terrible", and saurus, which means lizard.

As for the individual species, they were named in different ways. Often, dinosaurs are named for a distinctive characteristic. Baryonyx means heavy claw. Corythosaurus means helmet lizard. Tyrannosaurus means tyrant lizard — you get the point. Other times, they're named after the place where the first fossil was found, like Albertasaurus (from Alberta, Canada).

Before we start looking into individual species and their names, there are a few other Greek roots that can help you better understand dinosaur names.

- Draco: from **Rakon** – means Dragon;
- Hippos: from **Hippos** – means Horse;
- Hydro: from **Hydro** – means Water;
- Ortho: from **Orthos** – means Straight, Right, or Upright;
- Macro: from **Makros** – means Large;
- Micro: from **Mikros** – means Small;
- Mega: from **Megas** – means Huge;
- Morph: from **Morph** – means Shape;
- Poly: from **Polys** – means Many.

Now that we've got the boring stuff out of the way, let's finally look at some dinosaurs.

Tyrannosaurus Rex

Tyrannosaurus lives up to its reputation as one of the most fearsome animals of all time. Its powerful jaws had 60 teeth, each one up to 20cm (8 inches) long and its bite was around 3 times as powerful than that of a lion. Bite marks found

on Triceratops and Edmontosaurus fossil bones show that Tyrannosaurus could crunch through bone. Analysis of fossilised Tyrannosaurus dung show that it contained the bones of its prey.

The Tyrannosaurus skull was over 1.5m (5 feet) long and the cavity that housed the part of the brain responsible for smell was relatively large. Tyrannosaurus would have used its good sense of smell to hunt live prey and locate dead bodies to scavenge. It would have been able to scare off any other scavengers, so it didn't have to share.

Stegosaurus

one of the various plated dinosaurs

(   ) of the Late Jurassic Period (159 million to 144 million years ago) recognizable by its spiked tail and series of large triangular bony plates along the back. Stegosaurus usually grew to a length of about 6.5 metres (21 feet), but some reached 9 metres (30 feet). The skull and brain were very small for such a large animal. The forelimbs were much shorter than the hind limbs, which gave the back a characteristically arched appearance. The feet were short and broad.

Various hypotheses have attempted to explain the arrangement and use of the plates. Paleontologists had long thought that Stegosaurus had two parallel rows of plates, either staggered or paired, and that these afforded protection to the animal's backbone and spinal cord. However, new discoveries and reexamination of existing Stegosaurus specimens since the 1970s suggest that the plates alternated along the backbone, as no two plates from the same animal have exactly the same shape or size. Because the plates contained many blood vessels, the alternating placement appears consistent with a hypothesis of thermoregulation. This hypothesis proposes that the plates acted as radiators, releasing body heat to a cooler ambient environment;

conversely, the plates could also have collected heat by being faced toward the sun like living solar panels.

stegosaurus would have lived in family groups and herds, moving slowly through forests while eating the low-growing plants. Its front legs were considerably shorter than it's hind legs, making it adapted to nibbling the plants closest to the ground. Stegosaurus is the namesake for a large family of dinosaurs whose members were found all over the world. The Stegosaurus (its code name "Stego") was a peaceful herbivore and probably roamed the prehistoric highlands in herds that size from small to large numbers, grazing on low ground plants. An average Stegosaurus was about the size of an elephant, standing about 11 feet tall. It had a very low intelligence (its brain was the size of a kitten). The bony plates on it's back may have served a dual purpose of a body temperature regulation and protection from large flesh-eating predators. Different species varied in the number of plates on their backs.

Velociraptor

sickle-clawed dinosaur that flourished in central and eastern Asia during the Late Cretaceous Period (99 million to 65 million years ago). It is closely related to the North American Deinonychus of the Early Cretaceous in that both reptiles were dromaeosaurs. Both possessed an unusually large claw on each foot, as well as ossified tendon reinforcements in the tail that enabled them to maintain balance while striking and slashing at prey with one foot upraised. Velociraptor was smaller than Deinonychus, reaching a length of only 1.8 metres (6 feet) and perhaps weighing no more than 45 kg (100 pounds). Velociraptor appears to have been a swift, agile predator of small herbivores.

This dinosaur walked on two feet (bipedal) and could run very fast, perhaps up to 40 miles per hour. It had 80 very sharp teeth and sharp claws on its feet and hands. One of its claws on its feet was especially long and dangerous. This middle claw was up to 3 inches long and likely was used to tear into prey and deliver the killing blow.

The Velociraptor had one of the largest brains when compared to its size of any of the dinosaurs. It was likely one of the most intelligent dinosaurs.

Velociraptors were carnivorous, meaning they ate meat. They likely ate other plant eating dinosaurs and may have hunted in packs in order to bring down larger prey. One of the most famous fossils discovered involves a Velociraptor fighting a Protoceratops, which is a smaller plant eating dinosaur about the size of a large sheep

Triceratops

Triceratops was a rhinoceros-like dinosaur. It walked on four sturdy legs and had three horns on its face along with a large bony plate projecting from the back of its skull (a frill). One short horn above its parrot-like beak and two longer horns (over 3 feet or 1 m long) above its eyes probably provided protection from predators. The horns were possibly used in mating rivalry and rituals. It had a large skull, up to 10 feet (3 m) long, one of the largest skulls of any land animal ever discovered. Its head was nearly one-third as long as its body.

Triceratops hatched from eggs.

The Triceratops is one of the last known true dinosaurs, becoming extict 66 million years ago. Its name literally means "three-horned-face". Its sturdy, robust body meant not only that the Triceratops was not an easy prey — but also that many examples have been preserved as fossils, allowing paleontologists to study the species in relative detail. Its fossils are among the most common dinosaur fossils in the late Cretaceous.

Triceratops was about 30 feet long (9 m), 10 feet tall (3 m), and weighed up to 6-12 tons. It had a short, pointed tail, a bulky body, column-like legs with hoof-like claws, and a bony neck frill rimmed with bony bumps. It had a parrot-like beak, many cheek teeth, and powerful jaws.

Triceratops lived in the late Cretaceous period, about 72 to 65 million years ago, toward the end of the Mesozoic, the Age of Reptiles. It was among the last of the dinosaur species to evolve before the Cretaceous-Tertiary extinction 65 million years ago. Among the contemporaries of Triceratops were Tyrannosaurus rex (which probably preyed upon Triceratops), Ankylosaurus (an armored herbivore), Corythosaurus (a crested dinosaur), and Dryptosaurus (a meat-eating dinosaur).

Triceratops was a ceratopsian, whose intelligence (as measured by its relative brain to body weight, or EQ) was intermediate among the dinosaurs.

Many Triceratops fossils have been found, mostly in western Canada and the western United States. Paleontologist Othniel Marsh named Triceratops in 1889 - from a fossil found near Denver, Colorado, USA. At first this fossil was mistakenly identified as an extinct species of buffalo. The first Triceratops skull was found in 1888 by John Bell Hatcher. About about 50 Triceratops skulls and some partial skeletons have been found.

Spinosaurus

Spinosaurus is called "spiny lizard" because it had a series of large neural spines up to 6 feet (1.8 m) long coming out of its back vertebrae, probably forming a sail-like fin that may have helped in thermoregulation, mating rituals and/or intraspecies rivalry. Spinosaurus had a relatively flexible upper spine (these vertebrae had modified ball-and-socket-joints) so it could arch its back somewhat, perhaps being able to spread the sail (like opening the ribs of a fan).

Spinosaurus was bipedal (it walked on two legs). It was about 40-50 feet long (12-15 m) and weighed 4 tons or more (some paleontologists estimate it weighed up to perhaps 8 tons); it is the largest known spinosaurid (a type of large, meat-eating dinosaur). It had a large head with sharp, straight, non-serrated teeth in powerful, crocodile-like jaws. Its arms were smaller than its legs but were larger than the arms of most other theropods. It may have gone

on all fours at times. Spinosaurus may have been a good swimmer (it had paddle-like feet similar to those of water birds).

Spinosaurus walked on two muscular legs and was a relatively fast, two-legged runner. It may have gone on all fours at times, given its relatively long arms. Dinosaur speeds are estimated using their morphology (characteristics like leg length and estimated body mass) and fossilized trackways.

pinosaurus was a middle Cretaceous period saurischian ("lizard-hipped") dinosaur. It was a theropod, a tetanuran, and a spinosauroid (dinosaurs with elongated spines on their vertebrae from the Cretaceous period).

Allosaurus

Allosaurus was up to 38 feet long (12 m) and 16.5 feet tall (5 m). It weighed about 1400 kg. It had a 3 feet long (90 cm) skull with two short brow-horns and bony knobs and ridges above its eyes and on the top of the head. It had large, powerful jaws with long, sharp, serrated teeth 2 to 4 inches (5 to 10 cm)

long.

Allosaurus was a powerful predator that walked on two powerful legs, had a strong, S-shaped neck, and had vertebrae that were different from those of other dinosaurs (hence its name, the "different lizard"). It had a massive tail, a bulky body, and heavy bones. Its arms were short and had three-fingered hands with sharp claws that were up to 6 inches (15 cm) long.

Allosaurus was a huge carnivore, a meat eater equipped with sharp, pointed teeth in large, powerful jaws - it was the biggest meat-eater in its habitat. This theropod also had long, sharp clawed hands. Allosaurus probably ate large, plant-eating dinosaurs

Allosaurus was a large, fierce predator that could kill medium-sized sauropods (or sick or injured large sauropods like Apatosaurus) and many others of its contemporaries. An Apatosaurus (a large sauropod) vertebra was found with Allosaurus tooth marks on it. Allosaurus may also have been a scavenger.

Allosaurus may have faced competition from the meat-eating Ceratosaurus.

The first Allosaurus was discovered in 1877 in Colorado and named by Othniel Charles Marsh. This specimen offered just a few fragments of the dinosaur. Marsh went on to name other, seemingly unique dinosaur specimens, which, with time, were discovered to also be Allosaurus fossils, causing quite a bit of confusion in the early documentation of this dinosaur.

A more complete Allosaurus skeleton was discovered in 1879 by H.F. Hubbell, but was left unpacked at the time. In 1903, after Hubbell's death, the specimen was finally examined and turned out to be one of the most complete theropod skeletons unearthed to date.

In 1991, a joint team consisting of researchers from the Museum of the Rockies and the University of Wyoming Geological Museum found an Allosaurus fossil near Shell, Wyo. that was 95 percent intact — they dubbed it "Big Al." In 1996, the same team discovered "Big Al Two," the best-preserved Allosaurus skeleton to date.

Archaeopteryx

Archaeopteryx (meaning "ancient wing") is a very early prehistoric bird, dating from about 150 million years ago during the Jurassic period, when many dinosaurs lived. It is one of the oldest-known birds.

Archaeopteryx seemed to be part bird and part dinosaur. Unlike modern-day birds, it had teeth, three claws on each wing, a flat sternum (breastbone), belly ribs (gastralia), and a long, bony tail. Like modern-day birds, it had feathers, a lightly-built body with hollow bones, a wishbone (furcula) and reduced fingers. This crow-sized animal may have been able to fly, but not very far and not very well. Although it had

feathers and could fly, it had similarities to dinosaurs, including its teeth, skull, lack of a horny bill, and certain bone structures. Archaeopteryx had a wingspan of about 1.5 feet (0.5 m) and was about 1 foot (30 cm) long from beak to tail. It probably weighed from 11 to 18 ounces (300 to 500 grams).

Archaeopteryx was not a large creature. It was similar in size to a magpie, and the largest individuals might have reached the size of a crow.

Megalosaurus

was a very large theropod for its day, reaching between 23 and 30 feet (7 to 9 meters) long, and weighing 1.5 to 2 tons, it one of the largest predator in its area of occurrence. It had somewhat short, but strong arms with sharp, hooklike claws

on three fingers, perfectly designed for gripping onto prey and slashing at it. It also had long, powerful hindlegs, good for chasing down prey. Its tail, like most other theropods, was built to help balance it while moving. It had a long, narrow skull with sharp, bladelike teeth for slicing through the flesh of other creatures. But, his bite wasn't weaker than Allosaurus's, despite his skull was much more flattened. That's because he had a very thick, neck, helping him to bite the victim very strong. He's designed for run, and his tail was moving from side to side when this dinosaur is walking. Mentioned tail is very hard, helping Megalosaurus to balance. It presents half of the Megalosaurus's weight. His hunting techniques were very cruel. first, he taken prey on the groun, then eat the victim alive. It was a close relative of the larger

Torvosaurus of North America and the similarly-sized Edmarka, that nearly drove Allosaurus to outcompetition until considerably larger allosaurids such as Saurophaganax appeared. Megalosaurus for a long time was thought to have been a carnosaur, but closer analysis of the bones proved it was its own genus. The newer version of Megalosaurus looks much different than the old-fashioned version of dinosaurs, which looked like large, lumbering lizards instead of large, upright, birdlike reptiles. It used to be pictured looking closely like a large Komodo Dragon, walking on four stout, sprawled legs with its tail dragging on the ground.

Megalosaurus was a theropod dinosaur, whose intelligence (as measured by its relative brain to body weight, or EQ) was high among the dinosaurs. Megalosaurus was the first dinosaur fossil to be scientifically described and named. Buckland found Megalosaurus' fragmentary fossils in England in 1819. Megalosaurus was named in 1824 by William Buckland. In 1827, Gideon Mantell (not Ferdinand Ritgen) assigned it the scientific type species name, Megalosaurus bucklandii, honoring Buckland.

Megalosaurus trackways have also been found in limestone in southern England.

Diplodocus

Diplodocus was a long-necked, whip-tailed giant, measuring about 90 feet (27 m) long with a 26 foot (8 m) long neck and a 45 foot (14 m) long tail, but its head was less than 2 feet long. It was among the longest land animals ever. Its nostrils were at the top of its head and it had peg-like teeth, but only in the front of the jaws. Its front legs were shorter than its back legs, and all had elephant-like, five-toed feet. One toe on each foot had a thumb claw, probably for protection. A fossilized Diplodocus skin impression reveals that it had a row of spines running

down its back.

Diplodocus was more lightly built than the other giant sauropods, and may have weighed only about 10-20 tons. Its backbone had extra bones underneath it, which had bony protrusions running both forwards and backwards (anvil shaped), a "double beam", probably for support and extra mobility of its neck and tail. It may have used its whip-like tail for protection. A recent Diplodocus skin impression was found, showing a row of spines running down the back.

Diplodocus held its neck more-or-less horizontally (parallel to the ground). The long neck may have been used to poke into forests to get foliage that was otherwise unavailable to the huge, lumbering varieties of sauropods who could not venture into forests because of their size. Alternatively, the long neck may have enabled this sauropod to eat soft pteridophytes (horsetails, club mosses, and ferns). These soft-leaved plants live in wet areas, where sauropods couldn't venture, but perhaps the sauropod could stand on firm ground and browse in wetlands.

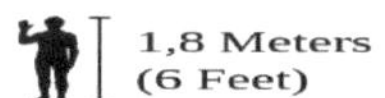

It used to be thought that the sauropods (like Diplodocus, Brachiosaurus and Apatosaurus) and Stegosaurus had a second brain. Paleontologists now think that what they thought was a second brain was just an enlargement in the spinal cord in the hip area. This enlargement was larger than the animal's tiny brain.

The first Diplodocus fossil was found by Earl Douglass and Samuel W. Williston in 1877 and was named by paleontologist Othniel C. Marsh in 1878. Many Diplodocus fossils have been found in the Rocky Mountains of the western USA (in Colorado, Montana, Utah, and Wyoming). Diplodocus means "Double-beamed."

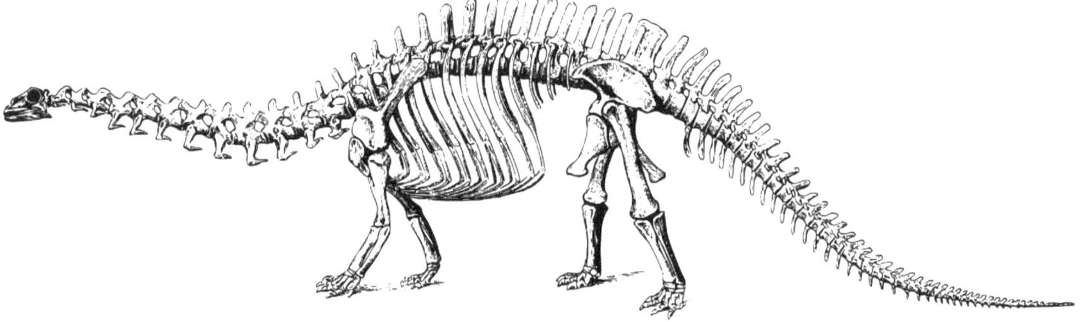

They likely had an avian respiratory system (which is far more efficient than a reptilian or mammalian one). Diplodocus would have spent its days browsing on trees, ferns, and bushes, from low levels to around 4 meters high. However, reconstructions suggest that it could also use its tail as a prop, giving it a stable tripodal posture (on its hind legs and tail), allowing it to reach up to 11 meters high. Like giraffes, it is believed that Diplodocus developed such a long neck as a feeding advantage.

Ankylosaurus

armoured ornithischian dinosaurs that lived 70 million to 66 million years ago in North America during the Late Cretaceous Period. Ankylosaurus is a genus belonging to a larger group (infraorder Ankylosauria) of related four-legged heavily armoured herbivorous dinosaurs that flourished throughout the Cretaceous Period (145.5 million to 66 million years ago).

Ankylosaurus was a huge armored dinosaur, measuring about 25-35 feet (7.5-10.7 m) long, 6 feet (1.8 m) wide and 4 feet (1.2 m) tall; it weighed roughly 3-4 tons.

Its entire top side was heavily protected from carnivores with thick, oval plates embedded (fused) in its leathery skin, 2 rows of spikes along its body, large horns that projected from the back of the head, and a club-like tail. It even had bony plates as protection for its eyes. Only its under-belly was unplated. Flipping it over was the only way to wound it.

Ankylosaurus had four short legs (the rear legs were larger than the front legs), a short neck, and a wide skull with a tiny brain.

Although Ankylosaurus has several striking features, its tail is probably the most interesting. Researchers believed it was actively used as a defensive weapon and was likely possible of crushing the bones of its would-be attackers — which, at the end of the Cretaceous, could have measured the likes of the T-Rex. Ankylosaurus' mouth suggests that it was an indiscriminate herbivore, feeding on whatever plants it could find lying around

Brachiosaurus

Brachiosaurus is another sauropod, like the Diplodocus. In 1903, paleontologist Elmer S. Riggs named the dinosaur Brachiosaurus altithorax — with the name Brachiosaurus being Greek for "arm lizard", and altithorax being Greek for "deep chest".

Like other Brachiosaurids, it had chisel-like teeth, its nostrils were on the top of its head, and it had large nasal openings indicating that it may have had a good sense of smell. Brachiosaurus had 26 teeth on its

top jaw and 26 on the bottom for a total of 52 teeth towards the front of the mouth.

The huge size of brachiosaurs led some researchers to suggest that they spent most of their time submerged in water, which would have served to buoy up their great weight. The location of the nasal openings—on top of the head and above the eyes—lent additional support to this idea. However, water pressure at the depths needed to cover these dinosaurs would have crushed their lungs and thus made breathing difficult or impossible. Other features of their skeleton show that brachiosaurs were well adapted to a life spent on land browsing the high treetops. Their skeletons were strong but not massive, so their weight could be supported without any help from water. Their great neckbones, for example, are so deeply excavated that they function as a lightweight framework of struts and plates.

Brachiosaurus and some of the other large sauropods (the huge long-necked plant-eaters) needed to have large, powerful hearts and very high blood pressure in order to pump blood up the long neck to the head and brain. The heads (and brains) of Brachiosaurus was held high (many meters) above its heart. This presents a problem in blood-flow engineering. In order to pump enough oxygenated blood to the head to operate Brachiosaurus' brain (even its tiny sauropod brain) would require a large, powerful heart, tremendously high blood pressure, and wide, muscular blood vessels with many valves (to prevent the backflow of blood). Brachiosaurus' blood pressure was probably over 400 mm Mercury, three or four times as high as ours.

Brachiosaurus was first found in the Grand River Valley, in western Colorado, USA, in 1900. This incomplete skeleton was described by paleontologist Elmer S. Riggs, who named Brachiosaurus in 1903. In 1909, Werner Janensch found many Brachiosaurus fossils in Tanzania, Africa. Many Brachiosaurus fossils have been found, in North America and Africa

Iguanodon

large herbivorous dinosaurs found as fossils from the Late Jurassic and Early Cretaceous periods (161.2 million to 99.6 million years ago) in a wide area of Europe, North Africa, North America, Australia, and Asia; a few have been found from Late Cretaceous deposits of Europe and southern Africa.

iguanodon was a plant-eating dinosaur that had a conical spike on each thumb.

guanodon was probably a herding animal, as evidenced by bonebed discoveries in Belgium. In these bonebeds, dozens of Iguanodon fossils were found together, suggesting that they congregated during their lives.

Iguanodon was named by Gideon A. Mantell in 1825; its teeth and a few bones were found in 1822 (perhaps by Gideon Mantell's wife, Mrs Mary Mantell) in Sussex, (southern) England. Mantell recognized the similarity between Iguanodon's tooth and that of the modern iguana, except the Iguanodon's was much larger. Iguanodon was the second dinosaur fossil named, and Mantell named it Iguanodon, meaning "iguana tooth." Hundreds of Iguanodon fossils have been found around the world, especially in Belgium, England, Germany, North Africa, and the USA. The type species, I. bernissartensis , was named by Boulenger and van Beneden in 1881.

Given its well-developed jaws, it's not clear what Iguanodon ate, although, given its size, it was probably a dominant herbivore. Remarkably, Iguanodon may have been bipedal in its early age, but became more quadrupedal as it got older and heavier.

Parasaurolophus

parasaurolophus (whose name means "near crested lizard") might not have the most famous name — but its look is definitely recognizable. Parasaurolophus was also a hadrosaurid, but its crest provides a notable difference to other

species.

Parasaurolophus grew to be about 40 ft (12 m) long and 8 feet (2.8 m) tall at the hips. It weighed about 2 tons. It had pebbly-textured skin, a spoon-shaped beak, and a pointy tail. It may have had webbed fingers, giving it a mitten-like hand, but some paleontologists argue that the web-like fossilized hands are an artifact of the fossilization process. Its sight and hearing were keen, but it had no natural defenses. It had a toothless, horny beak and numerous cheek teeth.

Parasaurolophus was an herbivore, eating pine needles, leaves, and twigs. Fossilized stomach contents have been found, consisting mostly of land plants.

Parasaurolophus was an ornithopod, whose intelligence (as measured by its relative brain to body weight, or EQ) was midway among the dinosaurs.

Parasaurolophus was described and named by Dr. William A. Parks in 1922 from an almost complete skeleton found in Alberta, Canada. Many fossils have been found in North America (in Alberta, Canada and New Mexico and Utah in the USA).

It's not clear what role the crest served. It may have been purely sexual display, or may have served for thermoregulation, or even acoustic resonance. Most likely, it served quite a combination of purposes, making it a unique feature, even in the diverse world of dinosaurs.

COELOPHYSIS

Coelophysis (pronounced SEE-low-FIE-sis) was a small, lightly-built dinosaur that walked on two long legs. This predator was about 9 feet long (2.8 m). It had light, hollow bones (hence its name), a long, pointed head with dozens of small, serrated teeth, three clawed fingers on its hands, and a long neck.

Two types of Coelophysis fossils have been found, 'robust' and 'gracile.' These two forms probably represent males and females.

Coelophysis most likely reproduced by laying eggs. Thousands of fossilized Coelophysis skeletons were found at the Ghost Ranch, New Mexico, USA. Many of them contained the bones of young Coelophysis in their abdomens - this evidence showed that Coelophysis either ate its young or gave birth to live young. It is more likely that they were cannibalistic (like many modern-day reptiles) since the tiny Coelophysis skeletons within the adult Coelophysis were not embryos, but young Coelophysis.

Coelophysis was discovered in 1881 by David Baldwin. It was named by US paleontologist Edward Drinker Cope in 1889. The type species is Coelophysis bauri (named by Cope and Colbert in 1964). There is some confusion about the naming of this genus arising from the fragmentary nature of the type specimen.

Thousands of Coelophysis fossils, including bonebeds (collections of many fossils of the same species in one location), have been found at Ghost Ranch, New Mexico, USA (these dinosaurs were called Rioarribasaurus at one time, but this name was later dropped). Several hundred Coelophysis fossil skeletons have been found in Arizona, New Mexico, and perhaps Utah. Both adults and juveniles have been found. There are two types of adults, "robust" and "gracile" - these two morphs may represent males and females.

Color

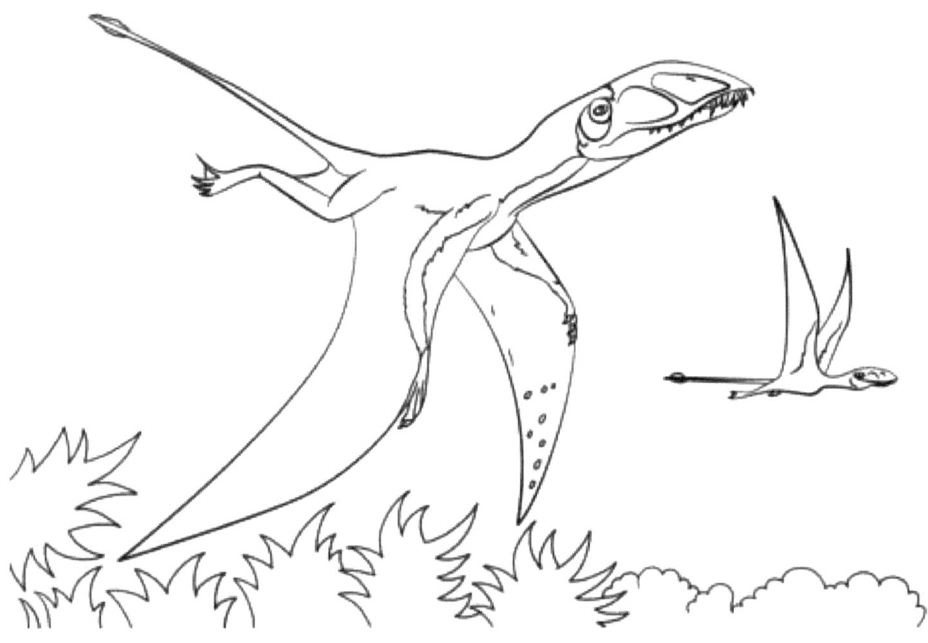

Draw Dinosaurs